Mixing Live Sound

An Application Guide
For The Audio Technician

A step by step guide for the everyday person wanting to learn how to mix for live bands and shows.

Mixing Live Sound

Mixing Live Sound

An Application Guide
For The Audio Technician

A step by step guide for the everyday person wanting to learn how to mix for live bands and shows.

Gregg J. Boonstra

ColRy Publishing

Mixing Live Sound

Mixing Live Sound (3rd Edition)

Published by ColRy Publishing
12101 North Rockwell Avenue
Box 256
Oklahoma City, Oklahoma 73162
A division of Vertical Vision, Inc.

Copyright ©2016 by Gregg J. Boonstra

All rights reserved. No part of this book may be reproduced or transmitted in any form or by any means, electronic or mechanical, including photocopying and recording, or by any information storage and retrieval system, without permission in writing from the publisher.

Printed in the United States of America

Mixing Live Sound

An application guide for the audio technician

A step by step guide for the everyday person wanting to learn how to mix for live bands and shows.

Mixing Live Sound

Contents

Introduction: ....- 9 -
Basic Systems (Flow)...- 12 -
Microphones: ....- 18 -
Instruments: How to capture their sound ....- 26 -
Monitors: Are they Necessary?...............................- 32 -
Do We Need So Many Buttons and Knobs?...........- 42 -
    Top of Strip/Gain Section....................................- 45 -
    EQ Section...- 48 -
    Aux Section ....- 52 -
    Bottom of Strip..- 57 -
    Master Section/Output Section............................- 62 -
How to Achieve a Great Mix ....- 66 -
Working Well With Others ....- 74 -
Contact the Author ....- 77 -
Notes: ....- 78 -

Introduction:

This book is written as a basic guide to help the everyday person serve as a great audio technician. The book has a specific flow. It has been laid out to first help you to understand the equipment involved and how to properly use the equipment. It is after you understand the proper use of the equipment that you can then use the equipment to mix the sound. This book is not all inclusive and does not cover every situation or piece of equipment you will encounter. This is written as a beginners guide helping you to understand the majority of equipment and how to best use it to make a great mix.

Some of the material in this book, you may consider elementary. Don't just skim over the parts that seem elementary to you. If you follow this book and the principles outlined, you will learn something and become a better audio engineer.

Almost anyone, including those with little to no technical or audio know how can become a great audio engineer by following the simple information and principles outlined in this book.

Take the time to read this book, maybe even while "playing" with your audio system. As you read, you can try the things we talk about and locate the parts of your audio system. By doing this, you will better connect the information in the book with the real life equipment you use.

Mixing Live Sound

Basic Systems (Flow)

All audio systems have a pretty common setup. Granted, some are more complex than others, but in general you will always find the same basic pattern or system flow. For this book, we will discuss a basic system using a standard setup.

The basic system flow is:
Source – Microphone or Direct Interface Box (DI) – Audio Mixer – Equalizer – Amplifier – Speaker

Source:
It all starts at the beginning. To even have a need for an audio system, you will have a source. A source is the original item creating the sound that you are attempting

to reproduce. The source may be a person, an instrument, or something else that generates the sound.

Microphone:
A microphone is the device used to capture the source in most circumstances. A microphone converts the sound waves into an electrical signal that can pass through the audio system

Interface Box (DI):
A direct interface box, commonly called a DI is a box that can connect to electronic sources such as guitars, keyboards, computers, etc. A DI will capture the electrical source and convert it for use in your system. The DI replaces the need for a microphone in a situation where you are using the DI.

Audio Mixer:
The audio mixer is the place where all the sounds come together. At the audio mixer there is usually a channel for each source and the ability to adjust the level

Mixing Live Sound

(loudness) of each source. At the mixer, each source can be fine tuned to blend with all the other sources and then routed back out of the mixer to be heard, recorded, and broadcast.

Equalizer:
An equalizer is used to adjust for system and room discrepancies. Every system and every room that you use an audio system in will sound different. The reason is that not all the frequencies of sound will be reproduced perfectly. The sound system might alter the frequencies of sound and each room will bounce and absorb sound waves differently. The equalizer will in a live sound system is used to make all sound frequencies equal in the environment. Sometimes in your home stereo you might use the equalizer to adjust for your personal preferences to sound such as adding more low end (bass) sounds. In a live sound system, these types of adjustments are best done at the audio mixer on the individual channels.

Amplifier:

An amplifier converts the electrical signal used in your audio system to an electrical signal that is used to make speakers work. Audio systems use small electrical currents to process audio. For audio to be turned back into something you and I can hear, it will need to be transformed into a much stronger electrical current to make the speakers work, that is what the amplifier does.

Speaker:

The speaker is the place where the sound is converted from electrical current back into a sound you can hear.

There are many shapes sizes and complexities of audio systems, but in general, this is the basic pattern that all systems will use.

Mixing Live Sound

Microphones:

There are many different types of microphones. In this chapter we will cover the most common types of microphones you will encounter and we will discuss some of their practical uses.

At the top of the microphone list there are two main categories. Wired and Wireless. Almost every microphone you use can be found in both a wired and wireless option. Years ago, most microphones were wired, but with the cost of wireless technology getting cheaper and the quality getting better, it is much more common today to see wireless microphones.

Wired microphones are simple and easy to use. You generally just plug them into a microphone cable and use it. Wireless microphones on the other hand require batteries, tuning, and can have interference. Wireless microphones are generally preferred when someone is going to be moving with it. By being wireless they do not need to drag a long wire with them giving them more freedom to move about. Wireless is also preferred because it has a neater and cleaner look. All the wires used to connect everything can become ugly and unsightly on a stage; therefore, by using a wireless microphone you will eliminate the ugly and unsightly wires on stage.

The next categories of microphones are Dynamic and Condenser. Both condenser and dynamic microphones are available as wired and wireless microphones. A dynamic microphone is a mic that simply operates on its own. A condenser mic requires a power source. In a wired system, the power source is called phantom power. Phantom power is commonly found at the audio console and is represented by a button or switch labeled +48V. By turning this switch to the on position, you will enable 48 volts of electricity to be sent down the microphone cable to whatever is attached at the other side. In addition to using phantom power some condenser mics have batteries to power them. In a wireless setup, the condenser mic will receive its power from the same batteries that power the wireless microphone, there is nothing you need to do to enable the power to a wireless condenser mic.

The basic difference between dynamic and condenser is simply one uses external power and one does not. It generally goes back to the design of the microphone and what the manufacturer chose to do to make the microphone work the way they intended it to. To add a little bit of depth to this discussion of dynamic verses condenser, generally a condenser microphone is more sensitive or more finely tuned. Usually the only reason you would choose to use one over the other is because you don't have the ability to use phantom power making it necessary that you use only dynamic microphones. If you have the ability to use phantom power, the choice of a microphones ability to perform for the intended purpose is more important than whether it is dynamic or condenser.

Now that we covered the main premise of microphones, we will now talk about several different styles of mics. We will not cover every style and every type ever made, but will cover the most common mics that you will use in a church setting.

The handheld microphone is the most common of all mics. They are used by vocalists and speakers as they present by holding them in their hand. Handhelds are also frequently cross used as instrument mics as there are some very good multi use hand held mics. To properly use a handheld microphone for a vocalist, the top of the mic should be directly in front of the vocalists mouth and about a thumbs length distance away from their mouth.

Headset microphone is a microphone worn on your head. These are typically used by preachers. There are a small variety of headset mics that are useful for vocalists, these are great because you don't need a microphone stand or your hands to hold the mic. Most headset mics have a short microphone boom on them. These are best positioned pointed at the corner of the person's mouth.

Lavalier or Lav microphone is a mic that is clipped to a persons tie or to their shirt. A lav is most frequently used by a preacher or someone who is speaking. A lav is not a good mic for a vocalist. The lav is best positioned between the second and third shirt button on a man's typical dress shirt. More technically, the lav is best placed about three inches below where your collar bones meet.

Instrument microphones come in all shapes and sizes. They usually have a very limited use because their design is for a very specific purpose. Positioning of an instrument mic varies by type of mic and the instrument being mic'd, but a good general rule is to place the head of the mic as close to the center of the sound source as possible.

Moving onto what are the best mics to use. There is no one simple answer. For vocalists, the best mic to use is typically a hand held microphone. A headset mic is great for a vocalist when you have the proper headset mic for a vocalist. Most headset style mics are not designed to be used for vocals. For speech or preaching, a headset mic is most preferable. When a headset mic is not available or not preferred by the presenter, the second best is a lav followed by a handheld mic. However, if the person speaking is a very dynamic presenter (loud to soft); you may want to lean more towards a handheld because of its ability to handle the louder side of a dynamic presenter. If you have a dynamic presenter and prefer to use a headset mic, make sure you use one that is able to handle the louder side. A mic that can be used for a vocalist is usually able to be used by a dynamic presenter. The best mic to use for an instrument is one that is able to process the sound properly. As I mentioned earlier, there are many varieties of mics that can be used for instruments.

Mixing Live Sound

Instruments: How to capture their sound

The topic of capturing the sound of an instrument is where a lot of division starts to occur between tech people and instrumentalists. We will discuss some of the many options that you will commonly use in a church environment, but this is not an all inclusive discussion of every possible way to capture the sound of an instrument. When planning the best way to capture the sound of an instrument both tech and instrumentalists must work together keeping in mind the end goal. In a church setting, the goal is not to have your way or what you know to be the best way. The goal is to present the best possible sound to the congregation while honoring and serving God in the process. Both the tech team and the instrumentalist must compromise, working together to present the best possible sound as a praise offering to God.

Keeping in mind that the best solution may be a compromise, as often as possible you will want to use a direct output of an instrument into a DI box to capture an instruments sound. By doing so, you will eliminate the need for a microphone and capture only the sound of the instrument. Using a microphone on an instrument opens more possibility of feedback and will not only pick up the sound of the instrument but any other sound that is in that area.

Not all instruments have an output option. There may also be good reasons for not using the output of an instrument that has an output option. In these two scenarios, you will need to use a mic to capture the sound.

One of the simplest and most consistent ways to mic a piano is to use a mic on a boom stand swung into the opening of the lid behind the built in music stand. Position the mic about half way left to right pointed behind where the piano hammers strike the strings. A piano is a very difficult instrument to mic. Also being in a live setting where there is so much other sound that you don't want getting into your piano mic makes it even harder.

To mic an acoustic guitar, you will want to use a straight stand holding a microphone pointed at the hole in the body of the guitar. You will want to make sure that the mic is not in the way of the person playing the guitar. To mic an electric guitar, you will use a mic pointed directly at the center of the speaker in the guitar amp. It is also important that the amp/speaker is pointed away from the audience. If possible, tilt the amp back at a 45 degree angle so that it is pointing up towards the ears of the person playing the guitar.

Drum kits are very loud and the sound leaks into every mic on stage. The smaller the stage and the smaller the room, the harder it is to keep the sound of the drums isolated. When possible you will want to use an electronic drum kit to eliminate all of the extra sound on stage. If an electronic kit is not an option, using a drum shield is a good way to reduce the amount of sound that spreads throughout the stage and the whole room. Drums cause a lot of controversy and are again an area where it is important to work together, compromise, and remember the end goal is more important than personal preferences. With all that being said, for the standard drum kit you will place two mics a few feet above the kit, one on each side. These mics are called overhead mics. They are used to capture the cymbals and chimes. For the toms and snare drums, you will want a mic close to each one of these pointed directly at the center of the head of the drum. There are some really great clamps made to clamp to the side of the drum that hold a mic right at the head of the drum. Make sure that you place this mic so that it is out of the way of the drum sticks. The kick drum you will mic from the front side of the drum kit. It is best to place the mic right at the opening of the hole in the kick drum. It is very important to use a mic that can handle the loud and low end sounds of the kick drum. Conga drums can use one mic at the top of the congas positioned between the two drums pointed down at the heads of the drums.

You will want to isolate each instruments sound as much as possible. If a mic is picking up the sound of two or three instruments, it will be hard to provide a good clean mix. It is very important that you properly mic everything. The quality of your mix can only be as good as the quality of the sound entering each and every microphone.

Monitors: Are they Necessary?

The short answer is yes stage monitors are necessary. It is important for the performers on the stage to hear themselves and the others that they need to be in tune with. Now with that being said, the key part is it is important for them to hear themselves and the others they need to be in tune/time with. I did not say they needed it loud or to have a full house mix on stage.

Monitors and monitoring solutions are another big area of disagreement and arguments between tech and performers in church. Just as we discussed in capturing the sound of instruments, monitoring is another area where tech and stage performers must compromise and work together keeping in mind the end goal is not about them, but about praising God in what you are doing and leading the congregation to do the same.

As a tech person, in ear monitors are always my preferred monitor solution. My reasoning is very simple. In ear monitors produce no noise or extra sounds on stage. This makes it easier to control stage volume and have a great mix for the audience. Not everyone is willing to use in ear monitors. More often than not, those who are unwilling to use in ears are unwilling to do so because they haven't given them a real try or have had a bad experience with them in the past.

To make in ear monitors work well, there are a few basic things that you need to know. If you are using a system where each person can control their own mix the first thing you need to know is that not everyone knows how to mix their monitors (even if they are the ones who tell you everything they know about mixing).

Mixing Live Sound

I have seen great musicians frustrated to no end because the in ears don't provide them what they want to hear. At least that's what they told me. With these people I simply plugged my headset into their mix and made the mix I thought they needed. Not surprisingly, they were very happy with the clarity and ability to hear everything once the mix was what they needed. When you have those who cannot mix for themselves you will need to spend the time to help them get their mix right and then fine tune it as they want changes made.

Another reason people don't like in ears is because they feel disconnected from the audience or simply they want to hear the audience. There are two simple ways to overcome this problem. You can simply put out a microphone in a position where it will primarily capture the sound from the audience and feed it to the in ear system. Putting out a mic like this is referred to as an audience reaction mic. In a multi channel in ear system I usually feed the audience reaction mic to the last channel of the in ear system. This makes it easy for the performers to find that channel and turn it up and down as they want to hear more or less of the audience. Another way to overcome the separation from the audience problem is to use earpieces that do not block out all the outside noise. Those cheap little foam covered ear pieces allow plenty of outside noise into a performer's ear. The problem with using a non blocking type earpiece is that they allow too much outside noise in making it hard to hear the in ear mix. Another problem to the non blocking type ear piece is that they are usually cheap ear pieces that do not adequately provide the audio response needed for the performer to hear the full range of the in ear mix.

In ears provide such great audio with amazing clarity that it can also become overwhelming. To help with this, you want to make sure it isn't too loud. When it is really loud, your ears start to shut down and you stop hearing what is being pumped into your ears. In addition to being too loud, you can divide the mix so it isn't as much for each ear to digest and process. I usually pan percussion and rhythm to the left ear and the harmony to the right ear. I also pan female vocals to the left ear and male vocals to the right ear. I don't do a hard pan, meaning I don't make it so that percussion and female vocals are only in the left ear, but I make them more prominent in the left and subtle/quieter in the right. By panning the different sound variations between the two ears, it makes it easier for your brain to process and interpret the different sounds. By simply panning the sounds, it will give the performer a clarity that they have never heard before. When you do use panning, don't pan everything. I leave the lead instrument and the lead vocalist centered or un-panned being equal in both ears.

There are times when in ear monitors are either not available or a performer is unwilling to use them. When in this situation you will use a speaker commonly called a wedge or a stage monitor to provide the performer with the audio they need to hear. When using a wedge, you want to use one that is sized appropriately. Wedges do not need to be huge, they need to provide clear sound to the person who is using it. You will also want to make sure the wedge is pointed at the ears of the person using it. A stage monitor pointed at someone's feet just adds more noise to the stage. No one hears through their feet.

When we talked about microphones we discussed the need to isolate the sound as much as possible. The same is true for stage monitors. You will want to isolate the sound from the stage monitor as much as possible to provide sound to the user but not "spill" any farther than it must. Each monitor makes more noise on stage leaking into every open mic and spilling off the stage into the audience. Using stage monitors quickly leads to a big mushy mess of sound on and off the stage.

Mixing monitors is very different than mixing for an audience. The goal isn't a nice pretty sounding mix. The goal is to provide the performer exactly what they need to hear to stay in sync with the other performers.

Mixing Live Sound

So what do they need to hear? There a few basic things that every good monitor mix must have. It must have the beat, usually provided by a percussion instrument. The mix must also have the leader. The mix will also include the other performers that the listener needs to stay in tune with. For a vocalist, you will want to include some of the other vocalists. You will also want to include some piano or other instrument that carries the tune. Finally the mix will include the person who is hearing that mix.

Not everyone needs to hear everything. The goal of the monitor is to provide the performer what they need to hear to stay in tune and in sync. I repeat this because it is very important. Adding too much to a monitor mix makes it harder, not easier for the performer to hear what they need. When using wedges, adding more to a mix also tends to add more noise to the stage and then creates the need to keep turning them up so the performer can hear creating more noise on stage and spilling into the audience making the mix muddy and unpleasant.

When a performer tells you they need more of something in their mix, the rough translation is that they cannot clearly hear what they are asking for more of. Sometimes the solution is as simple as turning up what they're asking for. However doing so may increase the noise levels and make it harder to hear. You should think about the request and possible solutions. If a performer is asking for more piano, the rough translation is that the performer cannot clearly hear the piano. Instead of instantly turning up the piano in their mix, ask yourself why cannot they not clearly hear the piano. Is their stage monitor positioned correctly so that they can clearly hear what is coming out of it? Is there too much of something else in their mix that if I turn down a few other things in the mix will it make the piano more clear and give them what they need? Every adjustment you make will affect something else. Just turning up the piano starts the vicious cycle of turning up everything else, creating more noise, creating the need to turn up more stuff, making even more noise, finally resulting in a muddy mushy mix that no one is happy with. When mixing monitors, you need to understand what each performer really needs to hear and find a way to provide them with a clean and clear mix of what they need without creating too much stage noise. In addition to you providing the performer with exactly what they need to perform, a performer must be willing to compromise and work with you and understand the goal is to help them perform and provide everyone especially the audience with the best sound possible.

Do We Need So Many Buttons and Knobs?

We will start now by looking at your audio mixer. Understanding that every mixer is different, you might have a few different buttons and knobs than we discuss here. We will cover the most commonly found buttons and knobs. If you have a digital audio console, the buttons and knobs we discuss might be available in a menu screen, but know for sure that they do exist. Understanding what each and every button does is important to using the audio mixer to make a great mix. When looking at the mixer, it may be intimidating. Hundreds of knobs and little buttons; how can you possibly know what each one of them does? There is a simple breakdown. Each channel has a row of buttons and knobs. Each channel has exactly the same buttons and knobs. If you learn one channel, you have learned almost the entire audio mixer. If you have stereo channels, they might be slightly different than the other channels, but usually pretty much the same, just a little more simplified. After input channels, you have an output section. Just like input channels, outputs are pretty much the same. So, don't look at the audio mixer as a big intimidating device. Look at the few different sections and realize that they are just copied over and over again. All you need to learn are a few things.

Each row of buttons and knobs flowing from top to bottom is called a channel strip. Each channel controls the audio flow for one input source. An input source might be an instrument, a vocalist, or a CD player. Whatever the input source is, the channel strip will control how the audio flows from that source.

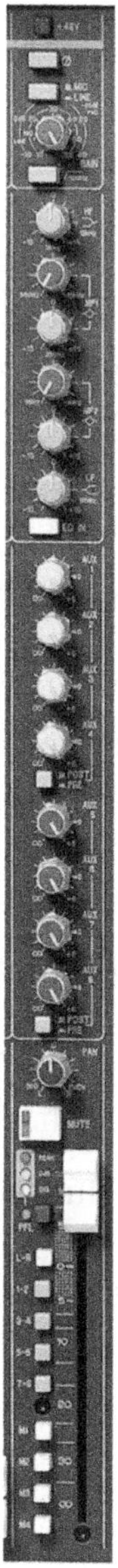

Mixing Live Sound

Top of Strip/Gain Section

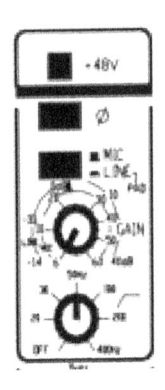

At the top of the channel strip you may see a button labeled +48v. This button is frequently red in color. The +48v button turns phantom power on and off. Phantom power is used for condenser microphones, and DI's that accept phantom power. If you are not using a device on that channel that requires phantom power, it is best to leave this in the off position.

Next you might see a button with a label that looks like a circle with a slash through it. This one is called phase reverse. What it technically does is switch the second and third wire on the microphone cable plugged into the console. You will rarely use this button, so we will not take the time to detail phasing.

You will also see a button labeled Mic/Line. The simplest way to explain this button is that it adjusts for the amount of sound coming into the audio mixer from the source. Typically you will use this in the Mic position. If you have a very strong signal coming from your audio source, you may want to try and switch this to the line position.

Frequently with a red top, there is a knob labeled Gain. This controls the amount of audio flowing down the channel strip from the source. Think of it as the faucet to the garden hose. If you turn it all the way up, you may have too much water coming into the hose. If you turn it all the way down it is as if you turned the hose off. Adjusting this knob will affect all the other functions of the channel. You typically want to set this one correctly at the beginning of rehearsal and then leave it be unless it is absolutely necessary to change it. This knob is easiest to adjust when you have a PFL meter for the channel. (We'll discuss PFL in a few minutes) You want to adjust the level of the gain so that

Mixing Live Sound

the audio in the channel makes the lights on the meter peak at about zero. Peak means to be at it's loudest. So when adjusting the gain for the piano, you want the lights to bounce up to the zero level without going over when the piano plays its loudest notes.

EQ Section

We just covered the most common things you will find at the very top of a channel strip. Next on the channel strip is the EQ section. The EQ section is a group of knobs that are often in two different colors. If you have six knobs in the EQ section, our knobs will be one color, and two of the knobs will be another color. Some audio mixers do not use six knobs, but may use three or four knobs. Each console may be slightly different, but most often you will find the EQ section to have a group of Six knobs. These knobs are broken down into frequency groups. There are typically four frequency groups. High Frequency (HF), High Mid Frequency, Low Mid Frequency, and Low Frequency (LF). To help understand these four frequency groups think of a piano keyboard. On one end of the keyboard you have the high notes and on the other you have the low notes. Divide that keyboard into four equal sections. The first section of the keyboard, the one with the high notes represents the high frequency. The next section of the keyboard represents the high notes of the middle section of the keyboard, the High Mid Frequencies. Following those first two sections you will find the lower notes of the middle section on the keyboard, these are the low mid frequencies. The final section of the keyboard are the low notes, this represents the low frequencies.

The first knob in the EQ section is the HF (High Frequency) turning this knob down will reduce the amount of the high frequency for that channel and turning it up will increase the amount of high frequency for the channel. When the knob is pointing straight up, it is neither adding nor reducing the amount of high frequency.

The next two knobs work together. They represent the High Mid Frequencies. The first of these knobs will select what frequency is being adjusted. This is called a sweepable EQ, where one knob will sweep across the frequencies. Picturing that piano keyboard, look at the section of the keyboard that represents the High Mid frequencies. Turning the sweep knob all the way to the left will make it so that you are adjusting the lower notes of the piano keyboard in that High Mid section. Turning the sweep knob to the right will move up that section of the keyboard until you are at the top note in the High Mid section of the keyboard. The second of these two knobs will control whether you are adding or reducing the High Mid frequencies selected by the sweep knob for the channel.

The next two knobs in the EQ section work together to control the Low Mid Frequencies for the channel. They work just like the two knobs for the High Mid frequencies, where the first knob is used to select the frequency, and the second knob is used to adjust whether you are adding or reducing the chosen frequency.

Mixing Live Sound

The last knob is the LF (Low Frequency). This knob will adjust whether you are adding or reducing the low sounds for that channel.

Depending on your audio console you may also have a button labeled EQ In or EQ On/Off. If this button is in the on position, the EQ section that we just talked about will be working. If this button is in the off position, the EQ section of the console will be turned off and the knobs we just discussed will do nothing.

Aux Section

The next section of knobs is the Auxiliary Send section. These knobs are labeled as Aux. You may have two aux sends (Aux 1 and Aux 2) or you may have twelve aux sends. The number of aux sends is not important. What they do is send the sound of that channel out of the corresponding output on the mixer. Let's say you have stage monitors that you mix from the audio console. You will typically use aux sends for the stage monitors. For our discussion, we will say you have two stage monitor mixes. Your two stage monitor mixes will be controlled by aux 1 (stage mix 1) and aux 2 (stage mix 2). We will say that all instrumentalists use stage mix 1 and all vocalists use stage mix 2. If an instrumentalist (stage mix 1) needs to hear more piano, you will find the channel for the piano and then on the piano channel you will find the aux 1 knob. When you find the aux 1 knob for the piano you will turn it up. By turning up the aux 1 knob on the piano channel you have added more piano to the stage mix 1 giving the instrumentalist more piano in their monitor. Like wise, if a vocalist (stage mix 2) needs to hear more piano, you turn up the aux 2 knob on the piano channel. This will now give the vocalist more piano in their monitor.

The most common use for auxiliary sends is to mix stage monitors. Another common use is for mixing a recording. In our scenario above, we have used Aux 1 and Aux 2 for stage monitors. If we wanted to make a recording, we might now choose to use Aux 3 for recording. In this case, if we want to hear more guitar in our recording, we will simply turn up the Aux 3 knob on the guitar channel. We will make adjustments to our recording by using the Aux 3 knob on every channel.

The other button you may find in your Aux section is a button labeled PRE/POST. There may be one PRE/POST button for the whole Aux section. There may be one PRE/POST button for each aux knob, or you may find one for Aux 1-4, and one for Aux 5-8, or some other configuration of PRE/POST buttons. Whatever the configuration is isn't important. What is important is that you use it correctly. What this button means is whether the sound going through the aux knob is pre fader or post fader. The fader is the slider at the bottom of the channel strip and is usually used to adjust the sound for that channel in the main room. We will talk more about the fader later. If you have the aux set to post fader, meaning after the fader, when you adjust the fader for more or less, you are also adjusting the audio going to the aux more or less. If you have the button set to pre fader, meaning before the fader it doesn't matter how you adjust the fader, the aux will not be affected.

Now let's apply this to our scenario above where Aux 1 and 2 represent stage mixes. If you have the aux buttons set to post fader meaning after the fader, any adjustments you make on the fader will also make adjustments to the stage mixes. This would not be good.

Mixing Live Sound

If you have piano in the stage mix so that the performers on stage can clearly hear the piano but then need turn down the amount of piano in the main room because it is too loud for your mix, you will turn down the fader. But, if the aux is post fader you will also be turning down the amount of piano going to the stage monitors making it harder for the performers to hear the piano. Stage mix aux sends should always be set to PRE fader. This way if you need to turn down the piano for the main mix you will not be changing the stage mix that the performers are relying on to stay in tune and in sync with one another.

In our scenario above we used Aux 3 for recording. This is great because we can make a great mix for recording separate from the way we have mixed for the main room. However, we can safely assume that any fine tuning we do to adjust the main room sound using the fader we will also want to make to the recording. In this case we will want Aux 3 to be POST fader. This way, when we turn down the piano for the main room we will also be turning it down for Aux 3 which is what is controlling the recording.

In short, an Aux used for stage monitors should always be PRE and an Aux for recording or broadcast should be POST.

Mixing Live Sound

Bottom of Strip

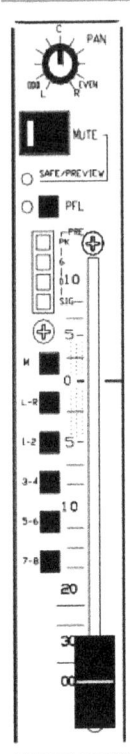

At the bottom of the channel strip you will find a knob labeled pan. When your audio system is setup as a stereo system (not all live systems are), the pan will move the sound from the channel from left to right. Also, for the groups (we will talk about groups in a minute) it will move the sound from odd number groups to even number groups. Group 1 is to the left and Group 2 is to the right. If the pan knob is pointed straight up, it is sending the sound equally to the left and the right.

Most audio consoles will have a button labeled mute or on/off. This button simply turns off the flow of audio through that channel. You will need to check on your specific console but typically the mute will turn off all the audio flowing through the channel. If muted it will not allow audio to flow through the aux sends and will not allow audio to the main room.

However, sometimes (it is rare) the mute button will only stop the audio from flowing to the Left and Right, while still allowing the audio to flow to the Aux sends.

Many consoles will have a meter for each channel. The meter for each channel is sometimes next to the fader and sometimes at the top of the console above the channel strip. I like when the meters are set PRE fade. This means that whenever there is audio in the channel you can see it on the meter. If the meter is set post fade, you will only see the meter lights when the fader is turned up. The meter will give you a good indication of how much audio is flowing through the channel and if you can't find where something is, if the meter is set to pre fade you can look for the meter that is moving on a channel.

There is also a button that is either labeled PFL or AFL. When you push this button it will send the audio from that channel to the headphone jack so that you can listen to it in your headphones. This is very helpful to make fine adjustments to a channel or to listen for a problem. PFL means Pre Fade Listen. With PFL whenever you press the button you will hear that channel in your headphones regardless of where the fader is adjusted to. AFL means After Fade Listen. With the AFL button pressed you will hear that channel in your headphones only if the fader is up and any adjustments you make to the fader will adjust how much audio you hear in your headphones from that channel.

You may see buttons labeled L-R, 1-2, 3-4, etc. These are groups. These buttons choose what groups to send the audio to. There are many different reasons for using or not using groups, but quite simply if your sound system is setup so that the main sound comes from the main Left and Right sends you will need to make sure L-R is pushed or turned on. There are some cases where you will send the audio to groups then to the Left and Right mains. In that situation, you will not press L-R, but you will press the appropriate group button.

Finally we come to the fader. The fader is the slider that moves up and down at the bottom of the console. It is the biggest of all the switches and knobs and is neither a switch nor a knob, but more of a slider type thing. You will slide the fader up and down to adjust the amount of that channel that you hear. For instance, if you want to hear more of the piano in the main room, you will push the fader for the piano up. If you want to hear less of the piano in the main room, you will slide the piano fader down.

We have just covered an input channel and all that it does. Everything we just talked about for one input channel is the same for every channel. You have just learned what 95% of the buttons and knobs on the audio mixer does.

Mixing Live Sound

Master Section/Output Section

The master section on every console is laid out differently. The master section is simply the last point for all audio before it leaves the console. For every Aux you will find a Master Aux knob. Even if you have Aux 1 turned up on a channel, the audio will not flow out of the audio console to the stage monitors if the Aux 1 master is not also turned up. There is also a master fader for Left and Right. In most audio systems, if the Left and Right Master sliders are not pushed up, no audio will flow to the main room sound system. You will see similar faders for each of the groups. You will also find a knob that controls the volume in your headphones. The master section is pretty simple and not an area of the console that you will interact with very often. You will spend the majority of your time focused on each individual channel. The master section is just the last stopping point for the audio before it flows on to its intended destination.

Mixing Live Sound

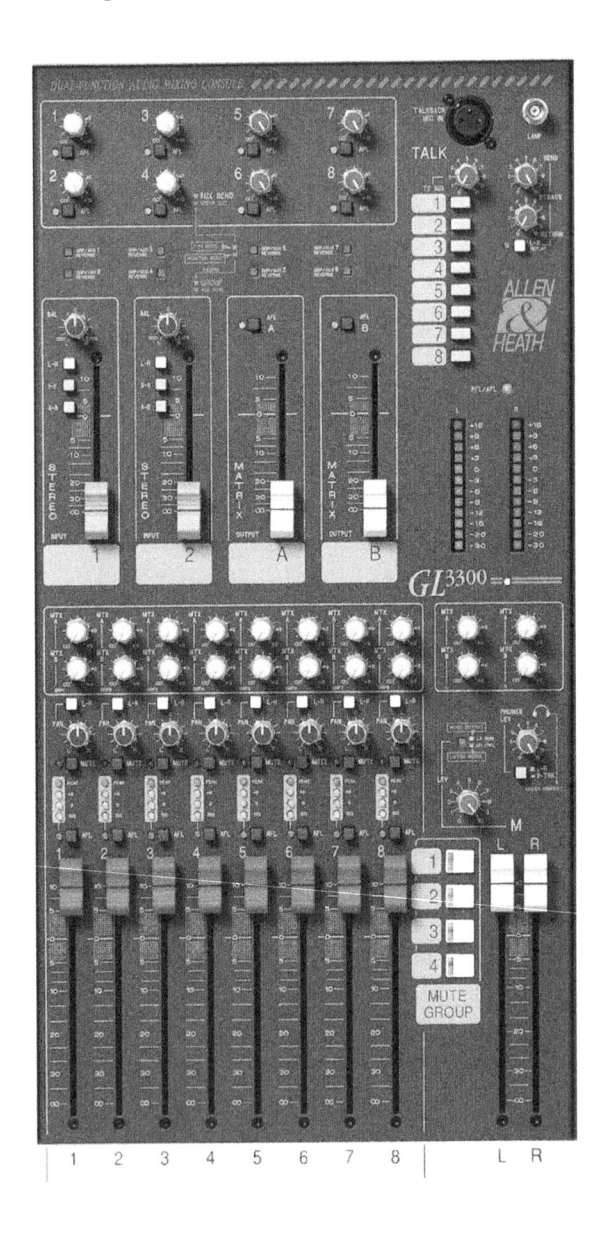

How to Achieve a Great Mix

We covered a lot of the foundation for mixing great audio. Before you start mixing, it is very important to know about the equipment, the buttons, the knobs, and how it all works. Without understanding how all the equipment works, you will be unable to use it well.

The first step to having a great sounding mix is to use the correct equipment. Use the correct microphone or DI for the source you are trying to capture. If you start with a poor sound at the beginning of the process, it will be hard to have good sound at the end.

When using a microphone, using the correct mic is vital. Just as important is the proper position of the mic. Make sure your microphone is in the best possible position to capture the sound. Make sure that your mic is as far away from other sound sources as possible. When setting up your stage, it is probably best to not put someone or something that needs a mic right next to the acoustic drum kit. The drum kit makes so much sound, that any open mic near it will pick up the drums and will just muddy the mix.

You cannot amplify what isn't there. If you give a mic to a quiet singer, it will be very hard to get any usable audio from that singer. This rule applies to every sound source. If the source is too quiet, you won't have anything usable to work with.

When you mix, you will need to think about the big picture. What is the overall goal? Remember, you are trying to achieve a complete sound, not just the sound of one instrument or vocalist.

You must keep control of all audio. You don't need to be overly controlling of what's happening, but you must maintain control of the sound. In maintaining control of sound, you will need to keep stage volume within reason and know when to say no. No to raising the volume too loud; no to making something stand out that should be blended in.

Stage monitors are the biggest offender to muddy mixes and sound that is out of control. Make sure that all amps and stage monitors are pointed away from the audience. Make sure that all amps and monitors are pointed at the ears of the person using it. A monitor pointed at someone's feet, just makes more noise move across the stage. Last time I checked, people do not hear very well with their feet. The goal for stage amps and monitors should be to keep them as quiet as possible while still giving the user what they need. Remember, stage mixes are not the same as the audience mix. The stage mix should be what a person needs to perform well with everyone else; no more, no less.

When working with vocalists, you will need to blend the vocalists together. Vocalists should not be blended against instruments. When working with the vocalist blending, ignore the instruments and listen specifically to the vocalists. Make sure they all balance against each other. The lead vocalists can stand out just a little bit from the other vocalists. When you have a soloist, you should make the soloist stand out above the other vocalists, but when the solo is over, make sure to bring that person back into the blend of the other vocalists.

Blend the instruments, just like in blending vocalists, instruments should be blended against each other. The final mix of instruments should be "under" the vocalists. The words being sung should be clear and understandable. If there is an instrument solo, you should bring up that instrument above the others, but when the solo is over, blend it back into the mix against the other instruments.

When mixing instruments, I suggest starting with the drum kit. An electronic drum kit will be easy. You want to set the gain on the kit where the PFL meters are peaking at 0. If you are working with an acoustic kit, start with the kick drum. Set the gain on the kick drum so that the loudest kicks peak the PFL meter at 0. Now blend in the tom mics followed by the snare and finally the overhead mics. If you have the capability, setup the drum kit mics to feed to a group so that you can easily adjust the whole overall volume of the drum kit with one slider instead of six or seven.

After the drum kit, blend into that the piano and keyboard. Next, bass guitar followed by electric guitars. Now move onto acoustic guitars followed by any remaining instruments. Sometimes it may be easier to move across your stage from right to left, that's fine, just use a pattern that works for you. I do suggest that you always start with the drum kit because it is very live and will be the hardest to get right and in balance with everything else.

When more means less. If your sound level is too loud it is neither enjoyed nor is it wanted. To get more, sometimes (often) you need to lower everything else. What I mean by this is, if you cannot hear the guitar, your natural response is to bring up the guitar. Just bringing up the guitar may not be your best option. You may need to lower everything else to make the guitar appear in the mix. When you are in the habit of bringing things up, you mix quickly gets too loud and out of control. You need to mix up to a comfortable level and not exceed that comfortable level. This rule applies to your mix for the audience as well as for stage. When someone on stage wants to hear more of something, learn to bring other things down so that what they want to hear stands out more.

Mixing Live Sound

If you can bring down the sound to the audience and notice little difference to the sound level, then the stage volume is too loud. You will need to work with the performers on stage to correct this problem. Stage volume polluting the audience area makes a very muddy sound that has no clarity and is not what is best for your audience.

If it sounds like everything is amplified, than everything is too loud. When mixed well, the sound should be natural sounding, not forced or pushed. Audio support should be unnoticed by the average audience member. Some obvious things to look for are: If people are covering their ears, it is too loud. If your ears ring during quiet spots, it is too loud. Remember, you are mixing for church, not a rock concert.

Your big picture goal should be to achieve a comfortable blend that seems to sound natural and full. The audio console is called a mixer. The term mixer describes the function. You are mixing the sounds together to achieve a nice blend. You are using a mixer, not a solo-er or louder-er.

Working Well With Others

Serving as an audio engineer should be a very rewarding and enjoyable experience. Unfortunately, all too often serving on a tech team is stressful and frustrating. The first step to reducing the stress is to form a trust with those on stage. If you don't work well with the performers and are second guessing all that they do, they will also doubt you and question your every move. This kind of distrust can quickly erode your ability to effectively serve together.

Although you enjoy serving and might be stressed, you need to focus. There is so much going on for the tech, that maintaining focus is difficult. Every detail matters, but not every detail needs your undivided attention nor should every detail take you away from focusing on achieving a great mix. It is your job to remove distractions and focus on mixing. I often see people wanting to hang out by the tech area and talk. When serving as a tech, it is a good idea to let people know that you will talk with them later and kindly ask them to hang out somewhere else. Right now you need to focus.

Lean on your team leader. Serving can be difficult. Someone is ultimately in charge. Know who this person is so that when you need them, you know who to go to. When you are feeling stressed or burdened by what you are doing, let the team leader know, it is their job to help you. If you are struggling with someone on your team or on the stage, let your team leader know. Don't lash out at the difficult person. Take a step back, and then let your leader know and ask your leader to help.

There may be times when someone disagrees with you or becomes angry. Instead of becoming angry in return, kindly ask that person to speak with your team leader. A good leader will hear what the angry person has to say and can then make a wise decision on how to handle the problem that arose.

Contact the Author

Gregg travels speaking to groups and organizations about technical and media systems. Offering training and insight to help them move forward technically. If you would like to schedule Gregg to speak to your group, you may contact him at: gregg.boonstra@servethroughmedia.com

Gregg also works with some clients directly one on one. This is done in person and via online video conferencing. If you would like to know more about working directly with Gregg, you may contact him by emailing: gregg.boonstra@servethroughmedia.com

Gregg is always happy to help you.

Mixing Live Sound

Notes:

Mixing Live Sound

Made in the USA
Middletown, DE
29 October 2018